Let's Go For A Trip Through the Solar System

Teach Me the Solar System

Copyright © 2012 Tamika K. Fordham

All rights reserved.

ISBN-13: 978-1974483822
ISBN-10: 1974483827

A Normal Size Star

The closest star to Earth is our Sun. It is at the center of our solar system. The Sun provides light and heat energy. Our Sun is an average size star. It appears bigger than other stars in the night sky, because it is much closer to Earth. The Sun is made up of hot gases, hydrogen, and helium. It's energy is produced in the core of the sun. Because heat and pressure comes from the inside of the Sun, it is much hotter near the core. We are able to have seasons, because the Earth is tilted on its axis and revolves around the Sun. Plants use energy from the Sun to produce food. We use its energy for heat, light, solar energy, and it gravitational pull.

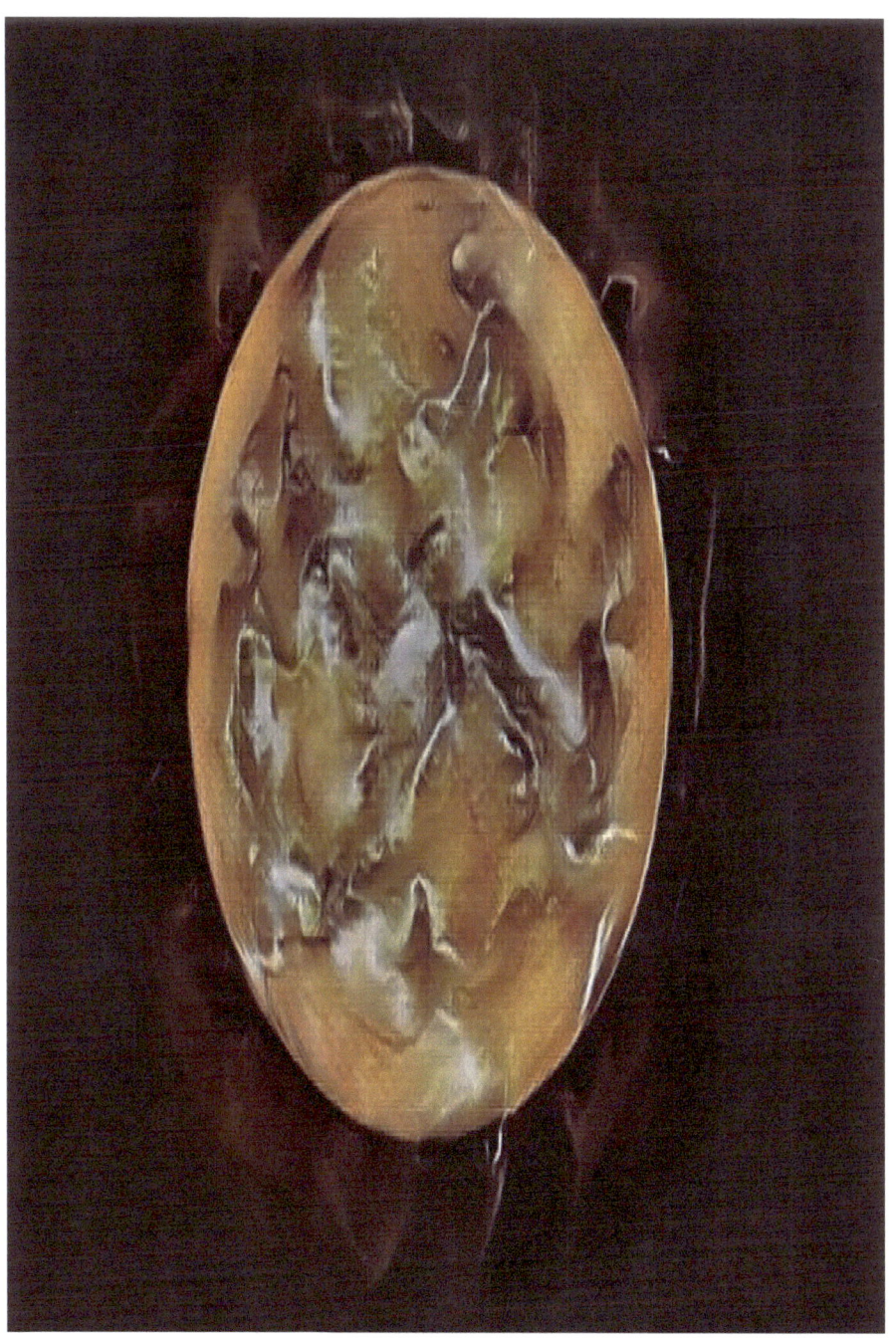

The Sun

The Crater Planet

Mercury is the closest rock planet to our Sun and the smallest planet in our solar system. Mercury is 36 million miles from the Sun. Mercury rotates slowly and has zero moons. One day on Mercury is 59 days on Earth. Its atmosphere is very thin. This causes Mercury's surface to be very hot during the day and extremely cold at night. Mercury is very close in size to our Moon. This planet is covered with rock dust. It is also covered with craters from meteors crashing into it!

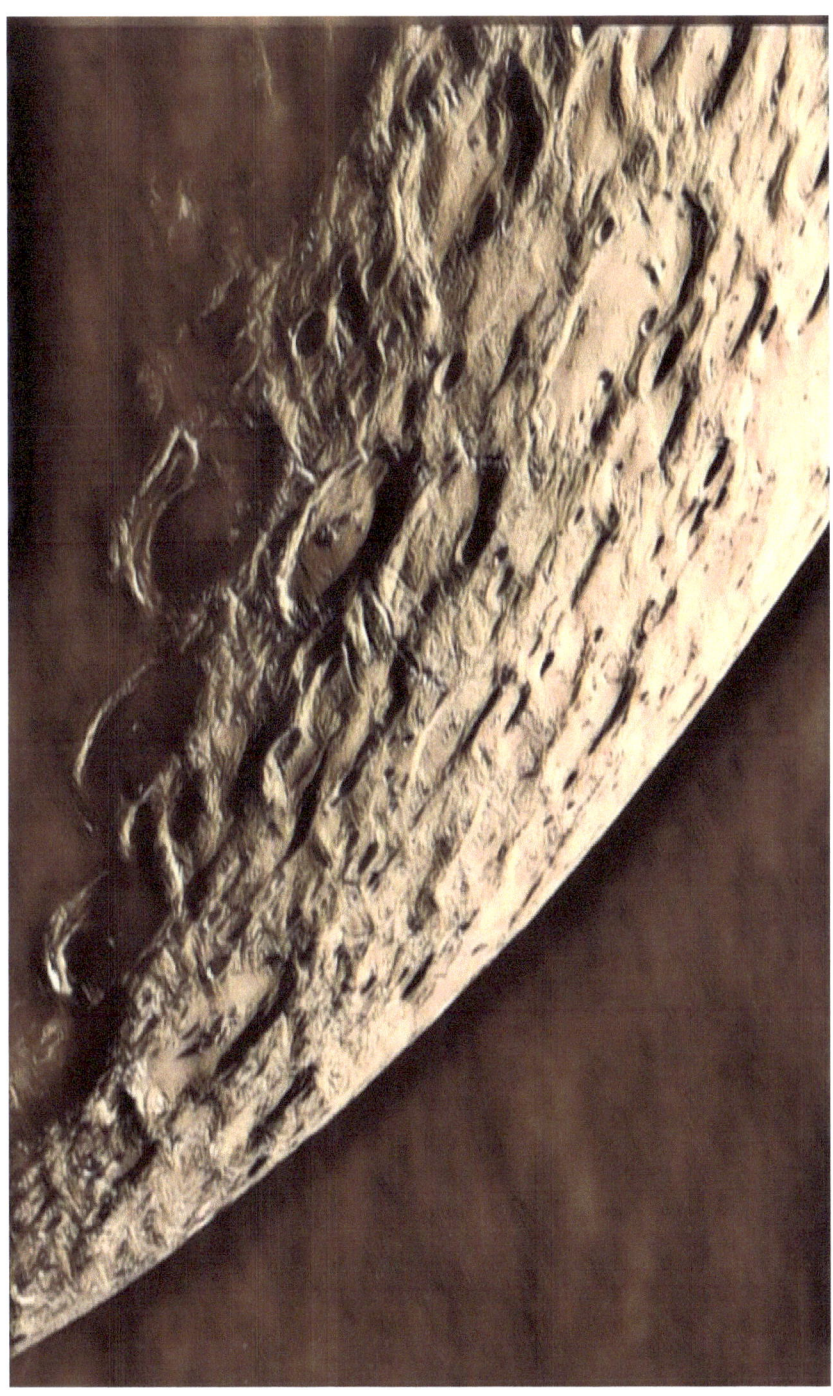

Mercury

The Sister Of Earth

Venus is the sister of planet Earth. These planets are very close in size. Venus is the second rock planet from the Sun. Venus does not have any moons. Venus has mountains, plains, and volcanoes flowing miles wide. It's atmosphere is very thick. It also has a thick cloud layer that traps the Sun's heat, making it extremely hot. That thick layer of clouds is made up of carbon dioxide. This makes the air very dense. The cloud helps reflect sunlight and looks very bright in the night sky. Venus is the hottest planet in our solar system. The air pressure is 100 times greater than on Earth. Because of the air pressure, heat absorption, and poisonous atmosphere it would be very hard to walk around on this planet. It would feel like walking through water in a swimming pool.

Venus

Planet Life

Earth is the third rock planet from the Sun. Earth is the only planet that we know of that is perfect for life. Here on Earth , you will find plants and animals of all kinds. Earth has the perfect amount of gases in its atmosphere that makes it suitable for life. Earth is 93,000,000 miles away from the Sun. It takes Earth 24 hours (1 day) to rotate. This allows Earth to have day and night. While, about 365 days (1 year) to orbit or revolve around the Sun. Earth is magnetic at its core and has north and south poles. Almost ¾ of Earth is made up of water. Earth's clouds protects its surface from heat from the Sun. Earth is tilted on its axis at 23.5 degrees. The highest point on this planet is Mount Everest. Earth has one moon. Because Earth is tilted on its axis, it gives us seasons as it revolves around the Sun.

Earth

The Red Planet

Mars is the fourth rock planet from the Sun. Mars is called "The Red Planet" because the iron in the soil gives it a reddish color. Because of the red dust, there are often dust storms on the surface. These dust storms can last for several months. Mars has two moons. It is the second smallest planet in our solar system. There are frozen ice caps on the poles of Mars.

Teach Me the Solar System

Mars

Shooting Stars

Almost any night, it is possible to see meteors as they fall to Earth. Sometimes so many fall at once that it is called a meteor shower. Meteoroids are small bits of rock and dust that are found in space. They are pulled into Earth's atmosphere by the force of gravity. Once a meteoroid enters the mesosphere layer of Earth's atmosphere, it begins to burn, creating a streak of light in the sky. They travel so fast that it gets very hot and glows. It is commonly called a falling or shooting star. Although many meteors burn completely before they reach the ground, not all of them do. If a meteor reaches the ground it is called a meteorite.

Asteroids are made of iron, nickel, and stone. Asteroids looks like small planets or small jagged rocks. They are located between Mars and Jupiter. There are thousands of Asteroids and over 5,000 of them are found between Mars and Jupiter in what is called the Asteroid Belt.

Asteroid Belt

The Storm Planet

Jupitar is the fifth planet from the Sun. It is the largest gas planet in our solar system. There is a Great Red Spot which is as large as Earth on Jupiter. This is a major storm that has continued to rage for hundreds of years. Jupiter is not a quiet planet. It has terrible storms and full of wind. Jupiter has more than 63 moons.

Jupitar

The Planet With The Most Rings

Saturn is the sixth gas planet from the Sun. Saturn has two wind systems operating in different directions. One system blows from east to west and the other blows from west to east. It's winds blow as high as 900 miles per hour in its atmosphere. Saturn has the most rings around its planet. It's rings are made up of hundreds of thousands of thinner rings. These rings were discovered by Galileo. The rings are composed of rock, ice particles, and dust. Jupiter is the only planet in our solar system that is bigger than Saturn.

Teach Me the Solar System

Saturn

The Sideways Planet

Uranus is the seventh gas planet from the Sun. Through a telescope Uranus appears bluish-green in color. Uranus is the only planet in our solar system that rotates sideways. Uranus does not spin like the other eight planets. Uranus rings are made of rocks, gases, and dust. Since its axis tips Uranus on its side, the rings appear to go around the top and bottom of the planet. Uranus takes 84 Earth years to orbit the Sun.

Uranus

The Blue Planet

Neptune is the eighth gas planet from the Sun. This planet is the smallest gas planet in our solar system. The planet appears blue because of the methane gas in its atmosphere. Like Uranus, Neptune also has a thick cloud cover. Scientist believe that Neptune is the coldest planet in our solar system.

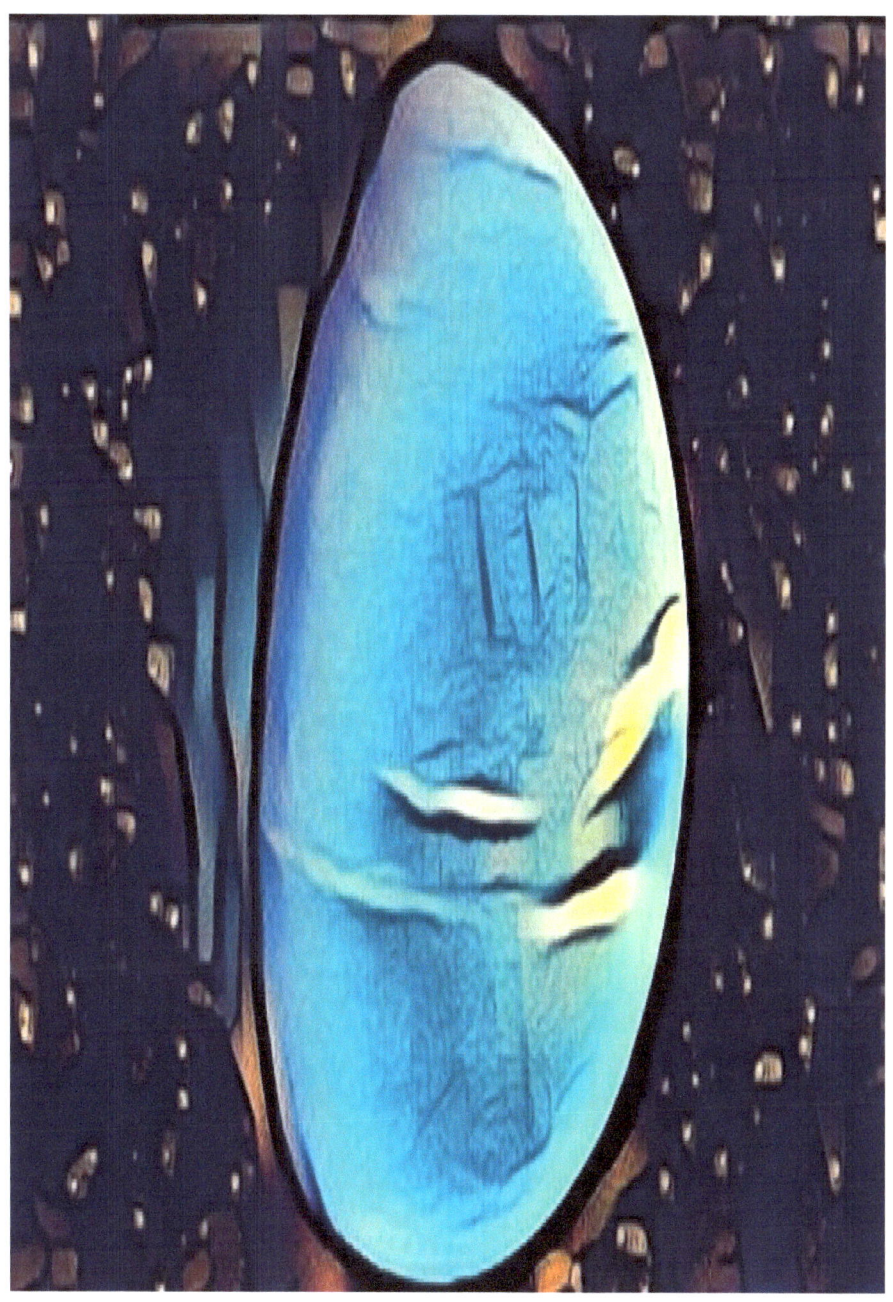

Neptune

ABOUT THE AUTHOR

    My name is Tamika K. Fordham, wife of Kendrick Fordham and mother of Kennedy Fordham. I was selected as 2016-2017 District Teacher of the Year for Calhoun County Public Schools and as 2017 STAR Teacher.

    I attended South Carolina State University and earned my Masters of Arts in Teaching in Early Childhood. I also attended Liberty University and received my Educational Specialist Degree in Leadership. I served in the United States Army Reserve throughout my college years and most of my teaching career. I became a highly qualified early childhood generalist teacher when I earned my National Board Certification.

Acknowledgements

My belief is a reflection of what my parents wanted for their children and of what I would want for my own children, as students in a classroom. My teaching style is best described as one that incorporates a diversity of teaching techniques, fun, energetic, and fair. I truly believe that learning takes place in a positive and safe environment filled with fascinating lessons, high expectations, and strategies that address each child's learning style.

I want to express my gratitude to my family, especially my mother (Eartha Carson), sister (Michelle Thompson) and husband (Kendrick Fordham), Dorchester County School District Four and Calhoun County Public Schools for providing me with the tools needed to be successful.

www.ingramcontent.com/pod-product-compliance
Lightning Source LLC
Chambersburg PA
CBHW041946240526

45473CB00033B/628